THE U.S. SPACE FORCE

THE CREATION OF THE U.S. SPACE FORCE

Martha Hubbard

Mitchell Lane
PUBLISHERS

2001 SW 31st Avenue
Hallandale, FL 33009
www.mitchelllanepub.com

First Edition, 2025.
Author: Martha Hubbard
Designer: Ed Morgan
Editor: Tammy Gagne

Series: The U.S. Space Force
Title: The Creation of the U.S. Space Force

Library bound: 979-8-89260-168-9
eBook ISBN: 979-8-89260-173-3

PHOTO CREDITS: p. 6, 8, 15, 17, 37 Shutterstock; p. 11, 19, 21, 23, 25, 27, 29, 35, 41 wikimedia; p. 31, 33, 39 spaceforce.mil; p. 42 NASA

CONTENTS

1 A New Kind of Warfare 4
2 The Founding of the U.S. Space Force 10
3 The History of the U.S. Space Force 18
4 U.S. Space Force Leaders 26
5 The Future of the U.S. Space Force 34

Find Out More 44
Glossary 45
Works Consulted 46
Index 48
About the Author 48

CHAPTER ONE

A NEW KIND OF WARFARE

A Russian satellite was being tracked by the U.S. military in 2019 when something extraordinary happened. The once single satellite split into two pieces. The first satellite seemed to produce a second. Sometime later, the two satellites floated uncomfortably close to a multibillion-dollar U.S. satellite. The Americans could not tell if the Russian satellites were a threat.

The United States expressed concerns to Russia, who pulled the satellite pair away peacefully. But several months later, a missile was fired into space from the second satellite. The possibility that this was an anti-satellite weapons test further concerned the U.S. military. Anti-satellite weapons are designed to cause serious damage to the satellites of rival countries. This appeared to be the first instance of such a test.

Early detection of an incoming attack can give ground forces a big advantage in avoiding harm.

INCOMING DANGER

In early 2020, American troops in Iraq were in danger. They didn't know it yet. But more than a dozen missiles had been launched from three bases in western Iran. These weapons, designed to strike from a distance, were headed straight for American ground forces. However, **infrared** sensors on U.S. satellites 22,000 miles (35,406 km) above Earth quickly detected them.

The satellites immediately sent data about the missiles to the Buckley Airforce Base in Aurora, Colorado. Air Force **Intelligence** tracked the missiles in real time. The agency warned the troops so they could seek cover in trenches or bomb shelters. Many troops suffered short-term brain injuries from the force of the impact. But there were no deaths. Modern space technology saved thousands of lives.

PROTECTING AMERICANS FROM SPACE

People around the world rely on satellites in many ways. Thanks to the development of the **Global Positioning System (GPS)**, the U.S. military has a better chance of keeping troops safe from enemy attacks. GPS satellites track enemy movements and warn leaders of dangerous threats. GPS technology also provides important information about weather patterns and deadly storms. This information helps forecasters keep people informed and safe. It also provides everyday people access to navigation technology, cell phone networks, and television broadcasts. Financial networks are also powered by satellites. People all over the world depend on satellites and GPS. This technology has created a lot of competition among the world's most powerful countries.

THE SPACE RACE

In the 1950s, the United States and the Soviet Union started the Space Race. The Soviets entered the record books first. They launched the first satellite, *Sputnik I*, on October 4, 1957. Space technology advanced quickly. On July 20, 1969, two Americans became the first human beings to set foot on the moon.

In the future, the setting for war will unlikely be limited to land, sea, and air. Space is the newest frontier. Many countries, including the United States, are trying to establish control in space. The race is on to develop the best technology and most powerful military high above Earth. The newest branch of the U.S. military was created for this purpose. On December 20, 2019, the National Defense Authorization Act was signed into law. This officially established the United States Space Force.

In 2020, a new partnership between the U.S. Space Force and the National Aeronautics and Space Administration (NASA) was announced. NASA is nonmilitary. Its projects focus on peaceful space exploration. But NASA understands that there are dangers due to unfriendly competitors. The Space Force and NASA now work together to develop and share new technologies. They share a goal of keeping space safe for all people who seek a better understanding of the world beyond Earth.

CHAPTER TWO

THE FOUNDING OF THE U.S. SPACE FORCE

The U.S. Space Force is the newest branch of the armed services. The last branch established before the Space Force was the Department of the Air Force in 1947. Although the Space Force is new, the idea of working to protect space is not. The Department of Defense worked for more than twenty years to establish a military space organization. A documentary created by the CSIS Aerospace Security Project discussed this goal. It stated, "The idea of warfare in space dates back to the beginning of the space age. After the launch of *Sputnik* in 1957, American and Soviet engineers went to work developing weapons that could defeat satellites."

According to the Space Force, its mission is to "secure our nation's interests in, to, and from space." The Space Force must protect U.S. satellites from nations that want to harm these devices. Everyday Americans rely on satellites to provide GPS navigation for their smartphones and automobiles. Satellite tracking of weather patterns can help farmers grow more crops and helps fishermen find waters with ideal temperatures for the fish they want to catch. Satellites also help utility companies manage power grids and help financial companies transfer money.

The American military depends heavily on GPS satellite technology. It helps all branches of the military track enemy activities. It also aids in fast global communications to help deal with a variety of threats. People in public safety depend on GPS technology to help them respond quickly and precisely to emergency calls. Satellites assist people living in rural communities by providing phone and internet coverage in areas that would otherwise have little or no service. This can make a lifesaving impact in the event of severe weather or a health crisis. Modern life depends heavily on working satellites.

Many people forget about satellites while using their cell phones or GPS systems, but these important objects orbiting Earth are what make this technology possible.

The Space Force falls under the Department of the Air Force, just as the Marines fall under the Department of the Navy. However, the Space Force has its own responsibilities. It is considered a separate branch of the armed services. The Air Force provides air support for ground forces. Its members fly over enemy territory to provide intelligence, surveillance, and **reconnaissance** (ISR). They may also take part in warfare through targeted airstrikes. The Space Force's mission is to organize, train, and equip its forces to protect U.S. interests in space. This includes satellite communications, missile warning systems, GPS navigation, and space **surveillance**.

The U.S. Space Force is its own military branch with its own set of responsibilities.

PUBLIC REACTION TO THE U.S. SPACE FORCE

President Donald Trump announced his interest in the creation of a space force in 2018. Public reaction was mixed. Late night television hosts and even some politicians openly joked about the idea. But Department of Defense members knew that protecting space was serious business. The department worked hard to push past people who made fun of the idea.

The Department of Defense took steps to create a detailed identity for the Space Force. A uniform was adopted, as well as the motto *Semper supra*. This means "always above" in Latin. A song reflecting the new motto was written and performed at public events. Space Force members were named guardians. Appealing advertising was created to attract new members. Officials created engaging websites and other content to show the public the importance of the Space Force. As its identity grew, public opinion began to change. The Space Force was becoming not only accepted but also respected.

THE SPACE FORCE ANTHEM

Every branch of the armed forces has its own song. This is a symbol of the branch's identity and pride. In 2022, the Space Force revealed its official song, "Semper Supra." The songwriting process began when the Space Force was formed in 2019 and took three years to perfect.

CHAPTER THREE

THE HISTORY OF THE U.S. SPACE FORCE

Flight has come a long way since the Wright Brothers created the first successful airplane in 1903.

The history of the Space Force began near the start of the twentieth century. The first successful airplane flight took place on December 17, 1903. Brothers Orville and Wilbur Wright worked hard to design the first functional airplane. Their hard work paid off. Orville flew their plane, the Wright Flyer, in Kitty Hawk, North Carolina. His first flight lasted twelve seconds. The Wright Flyer traveled just 120 feet (36 m). But this was a historic moment. The flight changed the course of the future in ways that could not be imagined in 1903.

Airplanes were first put to military use during World War I (1914–1918). They forever changed how wars were fought. **Aerial warfare** and ground attacks from the air gave an advantage to militaries. Airplane technology continued to advance. An interest in creating an independent air force emerged in the 1920s and 1930s. However, it wasn't until the United States' impressive use of air power during World War II (1939–1945) that the Air Force finally became its own military branch. On September 18, 1947, the Department of the Air Force was established as an agency of the Department of Defense.

Scientific advancements over the next two decades pushed explorers beyond Earth's atmosphere. In the 1950s, the Air Force developed rocket powered **ballistic** missiles. They could travel great distances and deliver explosives to specific targets. They could be launched from an aircraft or a ship. They could also be launched from rocket silos. These vertical structures also store weapons such as ballistic missiles.

The ability to drop bombs from the air helped the Allied Forces, which included the United States, win World War II.

The Air Force also got involved with satellites. But it was the U.S. Army that successfully launched the first U.S. satellite, *Explorer I*, in 1958. Space missions were carried out by all branches of the military at the time. In 1961, the Air Force was given responsibility for the U.S. military space effort.

The Air Force continued research and development of space technologies. But there were challenges. There was a need to separate the areas of air and space.

On September 1, 1982, the new Air Force Space Command was activated in a ceremony held at Peterson Air Force Base in Colorado. The command's motto stated that its members were the "guardians of the high frontier." General James V. Hartinger states in the documentary *War and Peace in the Nuclear Age*, "We activated the Air Force Space Command not to militarize space, but rather to preserve our right to self-defense and to **deter** space warfare."

Before the Air Force took over the space effort, the Army launched the first U.S. satellite, Explorer I, into space.

The new space command was focused on protecting satellites and other U.S. possessions in space. GPS technology was developed for military use. Soon, U.S. dependence on satellites was increasing. The Russians were regularly launching their own military satellites. Concerns about nuclear weapons were still high. A separate space command was needed to oversee U.S. space technologies to keep peace.

On December 20, 2019, the Air Force Space Command was formally renamed the United States Space Force. It was the first new branch of the armed forces created in seventy-three years. The Space Force was a product of a long history of air and space **innovation**.

CHOOSING A NAME

Space Force guardians were named in honor of the Air Force Space Command's guardians of the high frontier. The Space Force took name suggestions from citizens across the country. The other names submitted included *rocketeers*, *stormtroopers*, and *wookies*. However, the military branch thought the name *guardians* was the most fitting.

CHAPTER FOUR

U.S. SPACE FORCE LEADERS

The United States Space Command (SPACECOM) now operates under the U.S. Space Force. SPACECOM was established in August 2019. As stated on the SPACECOM website, "U.S. Space Command is responsible for delivering space capabilities to joint and combined forces and protecting and defending the space domain."

President Trump, remarking on the establishment of the U.S. Space Command, said, "From our nation's first days, America's military blazed the trails and crossed the frontiers that secured our nation's future. No **adversary** on Earth will ever match the awesome courage, skill, and might of American Armed Forces. Today, we salute the heroic men and women who will serve in SPACECOM and keep America's horizons forever bright and forever free." In December 2019, the U.S. Space Force became a reality. This was a historic day for Americans.

General John W. Raymond was made the first Chief of Space Operations. This is the Space Force's highest-ranking post. The Chief of Space Operations oversees the organization, training, and equipping of all space forces serving at home and overseas. Raymond became the first member of the Space Force. He had the very important role of building the Space Force from the ground up. And his job would not be easy. Many still did not take Space Force seriously.

When he was named the first Chief of Space Operations, General John W. Raymond also became the first member of the U.S. Space Force.

Raymond served as Chief of Space Operations for two years before retiring. In that time, he transferred many military members from Air Force space commands to the Space Force. He wanted the Space Force to have the best members possible. He managed the budget and set up three field commands.

General Raymond also built the Space Force's culture. He helped the new military branch overcome the media's jokes about it. Thomas Novelly of Military.com writes, "Raymond oversaw the creation of the service's uniforms, along with its motto, physical fitness standards, and official song." He even helped develop the new military branch's training. His leadership helped the Space Force start moving in the right direction from its beginning.

THE *GUARDIAN SPIRIT* HANDBOOK

Character, connection, commitment, and courage are a set of core values the U.S. Space Force calls the guardian spirit. *The Guardian Spirit Handbook* is a guide for guardians. It encourages them to apply these values to everyday daily life to become strong future leaders.

General B. Chance Saltzman is the Space Force's second and current Chief of Space Operations. He was formally sworn in on November 2, 2022. Saltzman continues to build upon General Raymond's achievements.

Saltzman faces many challenges as the branch's new leader. There is a need for updated equipment and systems. He must also deal with the problem of having more people interested in joining the Space Force than spots to fill. Saltzman must decide how large the Space Force should be. There are more than seventy countries operating in space. The Space Force must be ready for the dangers ahead by gathering the best people for jobs, improving equipment, and training guardians.

The year 2023 included a good deal of progress under Saltzman's leadership. Nine new units were activated. Space Force guardians were sent to Europe and Africa. International partnerships were established. The future depends on building strong networks and well-trained guardians.

General B. Chance Saltzman became the Space Force's second Chief of Space Operations in November 2022.

CHAPTER FIVE

THE FUTURE OF THE U.S. SPACE FORCE

The duties of Space Force guardians will change over time.

The Space Force is always changing to meet growing threats in space. Many discussions about the future are taking place. Improvements to how this military branch operates are being made. Advancements in guardian training are producing confident, combat-ready forces. Help from private industry is paving the way for better technology. New satellites are being launched to stay ahead of increasing threats. The future is uncertain. But the Space Force is quickly adapting to changing needs.

A new fleet of satellites will keep an eye on military targets anywhere in the world. It will be known as Ground Moving Target Indicator (GMTI). It will provide ground forces with video images almost in real time. The Space Force will be training guardians to operate these satellites. A fleet like GMTI "has never been done before," according to Brigadier General Devin Pepper. "Space-based targeting directly to the warfighter, directly to the shooters, is a game changer," he said during an Air Force Association conference in 2024.

On February 12, 2024, General Saltzman announced the creation of a new field command. This will be the Space Futures Command. It will consider whether the Space Force is developing technologies that will last over time.

Advancing technology will make it easier for the the U.S. military to watch targets from afar.

The Space Force uses a computer system to monitor satellites, spacecraft, and orbiting **space trash**. But this system was created in the 1980s. The Space Defense Operations Center (SPADOC) system is to be replaced with the Space Command and Control (Space C2) program. The new program includes many modern upgrades. It will be very effective against new cyber threats. However, Space C2 continues to be delayed. The tracking software it will use, the Advanced Tracking and Launch Analysis System (ATLAS), has had problems. This has pushed the replacement of SPADOC back by more than two years. The Space Futures Command wants to avoid these types of delays in the future.

The Space Futures Command will also weigh the future of space missions. There will be a **wargaming** center so it can experiment with new technologies. The goal is to stay ahead of new threats. Guardians serving in this field command will have new career paths to choose as training is updated. New educational and training opportunities will support guardians at all levels of service.

TRAINING FOR BATTLE

The Space Force held its first Red Skies exercise in December 2023. Guardians dealt with high-pressure situations that might occur in space in the future. Working as teams, they used a variety of strategies to protect the virtual satellites under threat. Officials hope to hold another exercise soon.

In 2020, NASA and the U.S. Space Force formed a partnership. While different, the two organizations share many goals. NASA's mission is the peaceful exploration of space. The Space Force's mission is to keep space safe so that peaceful exploration is possible. The two organizations will work closely together. They will share research and other data to achieve their common goals.

The NASA Artemis campaign has big plans. It aims to create a long-term U.S. presence on the moon in the future. Artemis also has plans to land humans on Mars.

Artemis is a character from Greek mythology. She is the goddess of the moon and the twin sister of the Greek god Apollo. Naming the new missions after Artemis reminds people of the Apollo missions that landed the first humans on the moon in 1969.

The U.S. Space Force will work closely with NASA on the Artemis project.

The Artemis program will take place across five separate missions. They have been named Artemis 1, Artemis 2, Artemis 3, Artemis 4, and Artemis 5. The first Artemis mission was completed in late 2022. It was an uncrewed test flight to the moon. The spacecraft then flew farther into space. It returned to Earth in a little more than 25 days.

Artemis 2 will have a human crew of four people. It will take them 5,523 miles (8,889 km) into space and around the moon before returning the crew to Earth. The mission is expected to take between eight and ten days. It is scheduled to launch no earlier than September 2025.

Artemis 3 will be the first space flight to land humans on the moon since 1972. This historic flight will land the first woman and the first person of color on the moon. The crew will spend one week on the moon before returning to Earth. The total time for the mission will be thirty days.

There will also be a space station built to orbit the moon. The station, named Gateway, will serve many purposes to help with the moon missions. Artemis 4 and Artemis 5 will deliver pieces of the space station. They will also transport more astronauts to the moon.

There may be even more Artemis missions. NASA has already awarded contracts for companies to build more rocket boosters. The safe exploration of space for the benefit of all humankind is the driving force behind both the Space Force and NASA's missions.

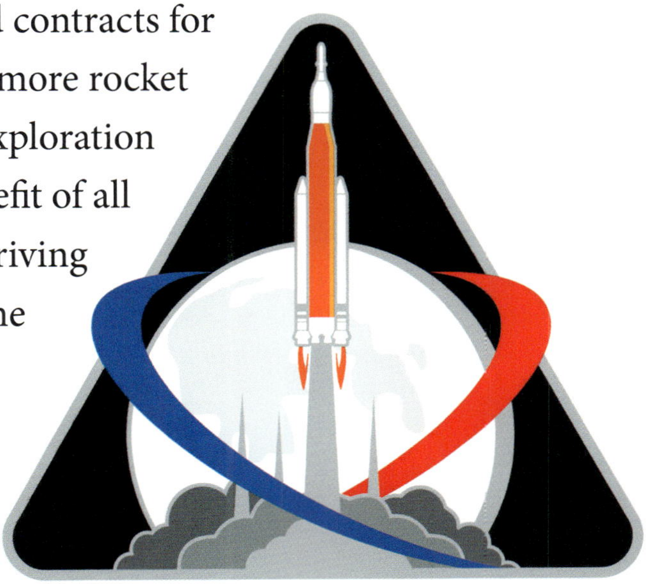

FIND OUT MORE

PRINT

Gagne, Tammy. *The Armed Forces Encyclopedia.* Minneapolis, MN: ABDO Reference, 2023.

Klatte, Kathleen. *Jobs in the U.S. Space Force*. New York: Rosen Publishing, 2023.

Sonneborn, Liz. *Space Robots*. Minneapolis: Lerner Publications, 2024.

ON THE INTERNET

NASA Space Place: NASA Science for Kids.
https://spaceplace.nasa.gov.

National Geographic Kids: Passport to Space.
https://kids.nationalgeographic.com/space.

Smithsonian Science Education Center: K-12 Space STEM Resources.
https://ssec.si.edu/fed-space-resources.

GLOSSARY

adversary
An enemy or opponent

aerial warfare
Military battles fought in the air

ballistic
Relating to projectiles or their flight

deter
To prevent or discourage from acting

Global Positioning System (GPS)
A network of satellites that provide location information to both civilians and the military

infrared
Low light waves that are invisible to the naked eye but can be detected with the help of technology

innovation
The use of creativity and knowledge to invent new technology

intelligence
Information gathered to assess threats to a person, a group of people, or a nation

reconnaissance
Military observation of enemy territory to gather important information

space trash
Pieces of rockets, satellites, or other technology that have been left in space

surveillance
Close watch or tracking of someone or something

wargaming
Computer gaming that simulates a military conflict in which the player must make critical decisions under pressure

WORKS CONSULTED

"1903 Wright Flyer," *National Air and Space Museum*, n.d. https://airandspace.si.edu/collection-objects/1903-wright-flyer/nasm_A19610048000.

"About Space Force." *United States Space Force*, n.d. www.spaceforce.mil/About-Us/About-Space-Force.

Brennan, Morgan. "Space Force Chief Says U.S. Is Facing a 'New Era' of Threats Beyond Earth." *CNBC*, April 20, 2023. www.cnbc.com/2023/04/20/space-force-chief-on-new-era-of-threats-beyond-earth.html.

Erwin, Sandra. "For U.S. Space Force, Raymond's Retirement Marks the End of an Era." *SpaceNews*, October 30, 2022. https://spacenews.com/for-u-s-space-force-raymonds-retirement-marks-the-end-of-an-era.

González, Jennifer. "U.S. Space Force: The Sixth Branch of the U.S. Armed Forces," *The Library of Congress*, December 20, 2022. https://blogs.loc.gov/law/2022/12/u-s-space-force-the-sixth-branch-of-the-u-s-armed-forces.

Hadley, Greg. "Red Skies: Space Force Launches New Orbital Warfare Exercise." *Air & Space Forces Magazine*, January 8, 2024. www.airandspaceforces.com/red-skies-space-force-new-orbital-warfare-exercise.

Howard, Brad and Morganteen, Jeff and Sheetz, Michael. "The Future of the U.S. Space Force." *CNBC*, June 3, 2022.
www.cnbc.com/video/2022/06/03/the-future-of-the-us-space-force.html.

Loverro, Douglas. "Why the United States Needs a Space Force." *SpaceNews*, June 25, 2018.
https://spacenews.com/why-the-united-states-needs-a-space-force/.

Novelly, Thomas. "Space Force Gets New Chief Space Officer for the First Time Since Forming in 2019." *Military.com*, November 2, 2022.
www.military.com/daily-news/2022/11/02/space-force-gets-new-chief-space-officer-first-time-forming-2019.html.

"President Donald J. Trump Is Establishing America's Space Force," *The White House*, February 19, 2019.
https://trumpwhitehouse.archives.gov/briefings-statements/president-trump-establishing-americas-space-force.

Scoles, Sarah. "NASA, Space Force Partnership Aims to Make Space Exploration Safe," *Science*, September 22, 2020.
www.science.org/content/article/nasa-space-force-partnership-aims-make-space-exploration-safe.

"United States Space Force History," *United States Space Force*, n.d.
www.spaceforce.mil/About-Us/About-Space-Force/History.

INDEX

Air Force, 6, 11, 14, 20, 22, 24, 30, 36
Air Force Space Command, 22, 24, 25, 30
anthem, 17
Artemis, 40, 42, 43
ballistic missiles, 20
Department of Defense, 11, 16, 20
Global Positioning System (GPS), 7, 12, 14, 24, 45
guardians, 16, 22, 25, 31, 32, 36, 38, 39
history, 19, 24, 25
motto, 16, 22, 30
National Aeronautics and Space Administration (NASA), 9, 40, 43, 44
National Defense Authorization Act, 9
public reaction, 16, 28
Raymond, John W. , 28, 30, 32
Red Skies exercise, 39
Saltzman, B. Chance, 32, 36
satellites, 5, 6, 7, 11, 12, 22, 24, 35, 36, 38, 39
Space Futures Command, 36, 38
Space Race, 8
training, 28, 30, 32, 35, 36, 38, 39
Trump, Donald, 16, 28
World War I, 20
World War II, 20, 48
Wright, Orville, 19
Wright, Wilbur, 19

ABOUT THE AUTHOR

Martha Hubbard is an Ohio author who loves to research and write about new and interesting topics. Following a career working in school libraries, writing for children was a natural next step. Researching the U.S. Space Force was especially rewarding because her father served in the Army Air Corps during World War II. His service and love of airplanes inspired her lifelong interest in aviation. Besides writing, her favorite activities include reading, spending time outdoors, and watching her favorite shows snuggled up next to her husband and their three dogs.